Plant Adaptations

D1489581

Check

📖 **Read the key points. When you finish, check the box.**

Key Points: What is an Adaptation?

What do the sharp spines of a cactus, the fuzz on a dandelion, and the stinky smell of a corpse flower have in common? They are all examples of plant adaptations.

An **adaptation** is a trait that helps a living thing survive in its environment. For example, the sharp spines of a cactus prevent the cactus from being eaten by animals. The fuzz on a dandelion helps the plant spread its seeds. Even the powerful odor of the Titan Arum, or corpse flower, helps the species survive. The rotten smell attracts beetles and flies, which pollinate the plant.

| cactus | dandelion | Titan Arum |

Adaptations are one reason why there are such a wide variety of plants and animals on Earth. Each species of plant or animal has its own set of adaptations.

✏️ **Complete the exercise.**

Test your knowledge

(1) Which of the following is the best definition of an adaptation?

A. It is a skill a living thing develops after birth.

B. It is how a living thing reproduces.

C. It is a trait that helps a living thing survive in its environment.

D. It is the way a living thing interacts with its environment.

Ans. ☐

(2) Which of these traits is least likely to be an adaptation in plants?

A. the spines on a cactus

B. the number of petals a flower has

C. the fuzz on a dandelion

D. the bright color of flowers to attract bees

Ans. ☐

Plant Adaptations

Think!

Think about where you live. What kind of plants do you see? Are they all the same?

 Read the key points below. When you finish, check the box.

Check ✓

Key Points: Adaptations and the Environment

Adaptations in plants are closely related to their **environment**, the area where a person, animal, or plant lives and grows.

For example, plants in the desert have many small leaves and thorns. This helps them to conserve water because there is almost no rain in the desert. They also have roots that grow near the surface of the soil to absorb rainwater quickly before it evaporates.

desert

Plants in the warm and wet environment of the rainforest grow leaves with waxy or shiny coatings that help them repel water. The coating is important because it prevents the leaves from growing mold which is not good for the plants.

rainforest

In the very cold and dry tundra, plants grow short and small in order to stay near the warm ground. Small leaves are also best for conserving water in an environment where there is very little rain or snow. These plants often have fuzzy stems to protect them from the cold wind.

tundra

✎ Complete the exercise.

Test your knowledge

What characteristics of plants can be seen as adaptations that are specific to the environment they grow in? Choose the correct answer from below.

A: desert   B: rainforest   C: tundra

(1) small leaves, fuzzy stems

Ans.

(2) many small leaves and thorns, roots near the surface of the soil

Ans.

(3) leaves with waxy or shiny coatings

Ans.

Chapter 1
Plant Adaptations

Think!

What are some other plant adaptations you can think of? Do you think all adaptations are visible to people?

 Read the key points below. When you finish, check the box.

Check ✓

Key Points: Seed Adaptations

When farmers or gardeners plant seeds, they are careful to plant them in a good spot. The young plants will need sunlight, water, and enough space to grow. But what happens to seeds in the wild, when there isn't anyone to plant them? This is where adaptations come in. Adaptations often allow plant **seeds** to travel and spread out. That way, at least some of the seeds are likely to end up in a good place to grow.

Here are some of the different ways that seeds can travel, thanks to their unique adaptations.

Animals: Plants have developed a few ways to move their seeds long distances with the help of animals. One example is a seed surrounded by a fruit, berry, or nut. When an animal eats the berry, the seeds in it will pass through the animal undigested. Then, they are released in the animal's droppings in another location. Other plants have developed a different way of using animals to move their seeds. Some seed pods have hooks or spines on the outside that get stuck to an animal's fur and are moved to a new location. Sweetgum seeds are an example of this adaptation.

Wind: Seeds of some plants have structures that catch the wind. Each dandelion seed is attached to a structure that works like a parachute. This structure is made up of about 100 tiny, feathery strands which help the seed get carried by the wind. Another example is how the two "wings" on a maple seed cause the falling seed to spin rapidly in the air. This spinning lets the seed travel farther than it would if it dropped straight down from the parent tree.

Exploding seed pod: Some plants have developed seed pods that "explode" when they are touched or when they dry out. This sends seeds shooting far away from the parent plant. Impatiens and geraniums are plants with this type of seed adaptation.

Water: Some plants produce seeds that float in water. Coconuts can travel long distances on the ocean from land to land. Some plants that tend to grow near streams, like foxgloves, also have this adaptation.

 Complete the exercise.

Test your knowledge

Determine how the seeds are transported to new places. Choose the correct answer from below.

A: wind   B: animals   C: exploding pods   D: water

(1) walnut

Ans.

(2) impatiens

Ans.

(3) maple tree seeds

Ans.

(4) blackberries

Ans.

(5) cotton

Ans.

(6) coconut

Ans.

Chapter 1
Plant Adaptations

✏️ **Use the word box below to fill in the blanks and review key vocabulary.**

Review the Key Points

An [_____] is a trait that helps a living thing survive in its environment. For example, the sharp spines of a cactus prevent the cactus from being eaten by animals. The fuzz on a dandelion helps the plant spread its seeds. Even the powerful odor of the Titan Arum, or corpse flower, helps the species survive. The rotten smell attracts beetles and flies, which pollinate the plant.

Adaptations are one reason why there are such a wide variety of plants and animals on Earth. Each species of plant or animal has its own set of adaptations.

Adaptations in plants are closely related to their [_____], the area where a person, animal, or plant lives and grows.

When farmers or gardeners plant [_____], they are careful to plant them in a good spot. The young plants will need sunlight, water, and enough space to grow. But what happens to seeds in the wild, when there isn't anyone to plant them? This is where adaptations come in. Adaptations often allow plant seeds to travel and spread out. That way, at least some of the seeds are likely to end up in a good place to grow.

Wind, [_____], exploding seed pods, and water are some of the different ways that seeds can travel, thanks to their unique adaptations.

> seeds / adaptation / animals / environment

🖩 **Complete the exercise.**

Math Mission

The amount of rainfall in an area can affect what type of adaptations a plant develops. The following graphs show rainfall in Place A and Place B.

(1) What is the amount of rainfall in March at Place A?

Ans. [_____] mm

(2) Which place has more rainfall in September?

Ans. [_____]

(3) In which place would plants adapted to a rainy environment grow better?

Ans. [_____]

Chapter 1
Plant Adaptations

💡 Read the mission. Then, answer the following questions to help you with your solution.

The Mission

Based on what you've learned about seed adaptations, imagine a new type of seed. The seed should have one or more adaptations that help it travel a long way from its parent plant.

Before you design...THINK!

1. Describe the mission in your own words.

2. Brainstorm your solution. Write your notes in the space below.

   Use the following questions to guide your thinking:

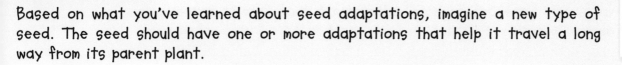

(1) What would your seed use for transportation: wind, animals, water, or some combination of these?

(2) In what kind of environment would your seed grow?

Chapter 1
Plant Adaptations

 Read the mission. Then, draw and evaluate your solution.

The Mission

Based on what you've learned about seed adaptations, imagine a new type of seed. The seed should have one or more adaptations that help it travel a long way from its parent plant.

Design   Draw or write about your solution below.

Evaluate

Have you ever seen a seed similar to the seed you imagined? If so, what kind of seed was it? If not, do you think a seed like the one you imagined could actually survive and grow? Why or why not?

Chapter 2
Pollination

Think!

What is pollination?

Check

📖 Read the key points. When you finish, check the box.

Key Points: What is Pollination?

Next time you see bees buzzing around a colorful flower garden, you might want to stop and take a closer look. These bees are hunting for food. The pollen and nectar from flowers are food sources for bees and their hives.

You might wonder why you should care about something as simple as a bee's next meal. It turns out, these buzzing bees have a big effect on you and other living things. That's because as bees are gathering food, they are also helping with an important process called pollination.

Pollination is the process through which plants produce new seeds. During pollination, pollen travels from the stamen of a flower to the pistil on the same or another flower. Without this process plants would not be able to **reproduce**, or make more plants.

Less new plants would create many problems for humans. We would not be able to grow as much food. We would also find ourselves in need of fabrics and the many other products that humans make from plants. There would not be enough new trees to replace fallen ones which means less trees to continue to recycle the air we breathe.

Bees are just one kind of animal that helps with pollination. An animal or insect that helps with pollination is called a **pollinator**. Sometimes pollination happens without a pollinator. No matter how it happens, pollination is something that we all depend on.

✏️ Complete the exercise.

Test your knowledge

(1) What is pollination?

A. When pollen travels from the stamen of a flower to the pistil on the same or another flower.

B. When pollen sticks to a bee or other insect.

C. When a flower blooms.

D. When a bee uses pollen to make honey.

Ans. ☐

(2) What is a pollinator?

A. A part of a flower.

B. An animal or insect that helps with pollination.

C. An animal that eats the whole flower.

D. A flower that only blooms at night.

Ans. ☐

Chapter 2
Pollination

Think!

How does pollination create new plants?

 Read the key points below. When you finish, check the box.

Check

Key Points: How Pollination Works

Flowers are not just pretty to look at. They are also complex structures. The different parts of a flower play different roles in pollination.

For example, the stamens of a flower produce a powdery material called **pollen**. Pollen grains are tiny and very light. Pollination happens when pollen grains reach another part of the flower called the pistil. Seeds then develop inside the flower—seeds that can someday grow into new plants.

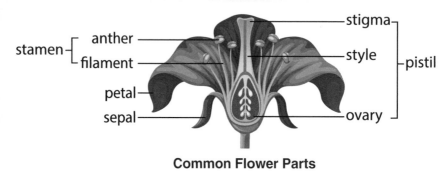

Common Flower Parts

Pollination can happen two different ways. First, when pollen travels from the stamen on a flower to the pistil on the same flower. Or, it can happen when pollen travels between these parts on different flowers with the help of pollinators.

 Complete the exercise.

Test your knowledge

(1) What is pollen?

A. a part of a flower that attracts bees

B. a part of the flower that grows seeds

C. a powdery grain particle produced by the stamen

D. a sticky substance that helps flowers grow

Ans. ☐

(2) Which part of the flower produces pollen?

A. the stigma

B. the pistil

C. the petal

D. the stamen

Ans. ☐

Chapter 2
Pollination

Think!

Do you think pollination happens more often with or without the help of pollinators?

Check

 Read the key points below. When you finish, check the box.

Key Points: Types of Pollinators

Insect pollinators like bees can help move pollen to where it needs to go. Of course, animals or insects do not do this on purpose. The pollen just sticks to a pollinator's body when it is gathering food from a flower. The pollen falls off later, pollinating other flowers the animal or insect visits.

Bees make great pollinators because their bodies are covered in tiny hairs that pick up a lot of pollen. Some other examples of pollinators are:

Butterflies: Butterflies have a long tongue that they use to drink nectar from flowers. Pollen sticks to their legs when they are taking a drink and they carry it to other flowers.

Hummingbirds: Long, tube-shaped flowers are pollinated by hummingbirds. The hummingbirds use their long beaks to reach deep inside the flower for nectar. Pollen then sticks to the hummingbird's head. It will carry pollen from one flower to another as it eats.

Bats: Bats often pollinate white or pale flowers that bloom at night. As with hummingbirds, pollen will collect on a bat's head as it feeds. It will then take the pollen to other flowers as it feeds.

 Complete the exercise.

Test your knowledge

Match the pollinator with the type of flower it is likely to pollinate.

(1) hummingbird

(2) bat

(3) butterfly

Chapter 2
Pollination

 Use the word box below to fill in the blanks and review key vocabulary.

Review the Key Points

[   ] is the process through which plants produce new seeds. During pollination, pollen travels from the stamen of a flower to the pistil on the same or another flower. Without this process plants would not be able to reproduce, or make more plants.

An animal or insect that helps with pollination is called a [   ]. Sometimes pollination happens without a pollinator. No matter how it happens, pollination is something that we all depend on.

Flowers are not just pretty to look at. They are also complex structures. The different parts of a flower play different roles in pollination.

For example, the stamens of a flower produce a powdery material called [   ]. Pollen grains are tiny and very light. Pollination happens when pollen grains reach another part of the flower called the pistil. Seeds then develop inside the flower—seeds that can someday grow into new plants.

Insect pollinators like bees can help move pollen to where it needs to go. Of course, animals or insects do not do this on purpose. The pollen just sticks to a pollinator's body when it is gathering food from a flower. The pollen falls off later, pollinating other flowers the animal or insect visits.

> pollinator / pollen / pollination

Complete the exercise.

Math Mission

The number of pollinators in an area can affect how much pollination takes place. Fewer pollinators can mean fewer seeds and fewer new plants. Answer the questions about pollinators.

(1) If one bee can pollinate 25 flowers an hour.
    How many flowers could 10 bees pollinate in an hour?

Ans. [   ] flowers

(2) One hummingbird can visit 1,000 flowers a day.
    How many flowers will a hummingbird visit in 5 days?

Ans. [   ] flowers

(3) If 12 butterflies take 2 hours to pollinate a field of flowers.
    How long would it take 4 butterflies to pollinate the same field?

Ans. [   ] hours

Chapter 2
Pollination

Read the mission. Then, answer the following questions to help you with your solution.

The Mission

Imagine you are an apple farmer. You have noticed that there are fewer bees in your apple orchard this year. You usually rely on the bees to pollinate the apple trees so they grow more apples each year. Design a device that can help you pollinate your apple trees.

Before you design...THINK!

1. Describe the mission in your own words.

2. Brainstorm your solution. Write your notes in the space below.

Use the following questions to guide your thinking:

(1) What traits does a bee have that make it a good pollinator?

(2) What are some traits of other pollinators that could be useful in your design?

(3) What materials could you use to build your device?

Pollination

 Read the mission. Then, draw and evaluate your solution.

The Mission

Imagine you are an apple farmer. You have noticed that there are fewer bees in your apple orchard this year. You usually rely on the bees to pollinate the apple trees so they grow more apples each year. Design a device that can help you pollinate your apple trees.

Design

Draw or write about your solution below.

Evaluate

Evaluate your design. What is one thing about your design that could be improved? How might you improve it? Is there a different pollinator you could have looked at for inspiration?

Chapter 3
Animal Adaptations

Think!

Are plants the only living things that develop adaptations?

Check ✓

📖 Read the key points. When you finish, check the box.

Key Points: Adaptations in Animals

Just like plants, animals can develop **adaptations** that help them survive in their habitat. Adaptations are traits that develop over time. If you look at animals in zoos and parks, you can easily see different types of animal adaptations.

There are two main types of adaptations in animals: one that helps an animal travel through its habitat and one that helps an animal get food in its habitat. This means an animal's body parts will be adapted to help it live in its habitat. As an example, let's focus on the feet of three different animals.

Polar bears have adapted to living on sea ice. Because they live in an extremely cold area, they have fur that grows in between the pads on their feet. These hairs not only prevent them from getting cold, but also give their feet a better grip on the ice.

Soil is the best habitat for moles. Soil has a stable humidity and temperature, and worms; a mole's favorite food. Moles can also protect themselves from predators by hiding underground. A mole's paws and claws help them dig tunnels in the soil and move around quickly.

Ducks live near bodies of water, such as rivers and ponds. A duck's feet are large and paddle shaped, so they can easily swim through the water.

✏️ Complete the exercise.

Test your knowledge

Match each adaptation to the habitat where each animal lives.

(1) river

●

● polar bear

(2) sea ice

●

● mole

(3) underground

●

● duck

Chapter 3
Animal Adaptations

 Think!

Can you think of other animals whose feet are well adapted to their habitat?

 Read the key points below. When you finish, check the box. Check

Key Points: How Adaptations Happen

Not all differences among living things are adaptations. Some are just that – differences. For example, members of the same plant species might have smooth or rough seeds. Or, members of the same animal species might have different color eyes or fur. These traits are passed from parent to child, even though they might not have a clear impact on an animal's survival.

But sometimes, a particular trait does turn out to be helpful for survival. That is when it can become an adaptation. Here is an example of the process where a trait becomes an adaptation.

1. A bird with an especially strong beak hatches from its egg. This bird has an easier time cracking open seeds to eat than other birds nearby.
2. There is a food shortage. Many of the birds do not survive. But the bird with the strong beak survives because it can get food more easily.
3. The surviving bird has babies. The baby birds inherit the same strong beak trait.
4. The process repeats many times.
5. Over time, the strong beak becomes common in the bird population. The strong beak is an example of an adaptation.

 Complete the exercise.

Test your knowledge

Write the letters in the correct order to show how a trait becomes an adaptation.

A. A population of lizards, some that are dark brown and some that are light brown, moves to a new area where the ground is tan and sandy.
B. Light brown lizards reproduce and pass their skin color on to their offspring.
C. The light brown skin becomes an adaptation for the lizards.
D. After several years, there are more light brown lizards and only a few dark brown lizards.
E. More dark brown lizards are eaten than light brown lizards, because they are easier for predators to see on the sandy ground.

Ans. $A \rightarrow \quad \rightarrow \quad \rightarrow \quad \rightarrow$

Chapter 3
Animal Adaptations

Think!

What type of adaptation is a bird's beak? Does it help the bird get food or move around in its habitat?

Check

📖 Read the key points below. When you finish, check the box.

Key Points: Adaptations in Birds

Even a single species of animal can have a broad range of varied adaptations. Bird beaks are a great example of this. You can see how bird beaks are suited to many different environments and types of food.

Hummingbirds: Hummingbirds have very long bills, almost like straws, to reach the nectar in flowers.

Herons: Herons and other birds that wade in the water use long, pointed beaks to grab or even stab at small fish or frogs.

Cardinals: Cardinals and birds that eat seeds typically have a strong, cone-shaped beaks, for breaking open seeds.

Hawks: Hawks and other birds of prey have hook-shaped beaks to help them catch and eat small animals.

✏️ Complete the exercise.

Test your knowledge

Match the bird to what it eats based on its beak.

(1) fish ●   ● hummingbirds

(2) small animals ●   ● hawks

(3) seeds  ●   ● herons

(4) flower nectar ●   ● cardinals

Chapter 3
Animal Adaptations

  Use the word box below to fill in the blanks and review key vocabulary.

Review the Key Points

Just like plants, animals can develop [ ] that help them survive in their habitat. Adaptations are traits that develop over time. If you look at animals in zoos and parks, you can easily see different types of animal adaptations.

There are two main types of adaptations in animals: one that helps an animal travel through its habitat and one that helps an animal get [ ] in its habitat. This means an animal's body parts will be adapted to help it live in its habitat.

Not all differences among living things are adaptations. Some are just that – differences. For example, members of the same plant species might have smooth or rough seeds. Or, members of the same animal species might have different color eyes or fur. These traits are passed from parent to [ ], even though they might not have a clear impact on an animal's survival.

Sometimes, a particular trait does turn out to be helpful for survival. That is when it can become an adaptation.

Even a single species of animal can have a broad range of varied adaptations. Bird [ ] are a great example of this. You can see how bird beaks are suited to many different environments and types of food.

food / child / adaptations / beaks

Complete the exercise.

Math Mission

The table on the right shows the results of an experiment to determine how much water, small marshmallows, rubber bands, and toothpicks can be obtained using the four tools which represent a bird's beak in 10 seconds.

		Food type			
		Water	Small marshmallows	Rubber bands	Toothpicks
Beak type	Scissors	0 ml	4	3	6
	Spoon	25 ml	6	12	2
	Tweezer	0 ml	2	16	2
	Binder clip	0 ml	3	24	10

(1) How many small marshmallows can you get with scissors?

Ans. [marshmallows]

(2) How many toothpicks can you get with binder clips?

Ans. [toothpicks]

(3) Which tool is the best way to get water?

Ans. [ ]

Chapter 3
Animal Adaptations

💡 Read the mission. Then, answer the following questions to help you with your solution.

The Mission

Study the image and text to the right. Use what you've learned about adaptations to imagine an animal that would live in the environment shown in the image. The animal could be one you've seen before, or one from your imagination.

Before you design...THINK!

1. Describe the mission in your own words.

This environment gets a good amount of rain each year. It stays damp and dark, so the ground is soft, muddy, and full of puddles. Other living things in this environment include many small insects, bats, and lizards.

2. Brainstorm your solution. Write your notes in the space below.

Use the following questions to guide your thinking:

(1) What do you think the weather is like in the location shown?
(2) What kinds of plants do you think would grow in this location?
(3) What kinds of adaptations might help an animal survive in this kind of environment?

Chapter 3
Animal Adaptations

Read the mission. Then, draw and evaluate your solution.

The Mission

Use what you've learned about adaptations to imagine an animal that would live in the environment shown in the image. The animal could be one you've seen before, or one from your imagination.

Design

Draw or write about your solution below.

Evaluate

Review your design. Are there any adaptations you might have also included? Are there any that might be unhelpful to the animal?

Chapter 4
Habitat Change

Think!

What are some features of your habitat?

Check

 Read the key points. When you finish, check the box.

Key Points: What is a Habitat?

Every living thing is suited to a particular environment: its **habitat**. A habitat is the natural home or environment of an animal, plant, or other living thing. A species' habitat includes other kinds of plants and animals that live in the same environment. It also includes things that are not alive such as soil, air, and water.

Changes to a habitat often happen slowly. For example, each year the temperature might rise a little each year or the amount of rainfall might drop.

Sometimes, when a habitat changes, it is not a big deal. Think about when the temperature drops in the winter. Just as you can put on a hat and scarf, other living things have ways of handling the change. For example, mammals often grow thick fur in the winter to help keep them warm. Remember, you learned in Chapters 1 and 3 that plants and animals have the ability to adapt to their habitats.

Complete the exercise.

Test your knowledge

Match the best habitats for the animals in each photo.

(1) deer

●

● freshwater river habitat

(2) bearded seal

●

● arctic habitat

(3) catfish

●

● temperate forest habitat

 Think!

Chapter 4
Habitat Change

How do you adapt to changes in your habitat?

 Check

📖 **Read the key points below. When you finish, check the box.**

Key Points: Problems with Changing Habitats

As you learned on the previous page, even if a habitat changes, species can sometimes adapt. Sugar maples, the trees that produce maple syrup, are an example of this. These trees grow in the northeastern United States and Canada. Temperatures in this region have been gradually rising, so now the area where these trees grow is shifting further north, where it's cooler.

If changes to a habitat are extreme or happen quickly, plants and animals may not have enough time to adapt. Members of the species might have to move somewhere else, or, over time, they might adapt to the new conditions. If one of these things doesn't happen, the species could be come **extinct** or die out.

For example, after a forest fire, many animals have to go elsewhere to look for food and shelter. The plants and trees they depend on have been damaged or destroyed. The area will generally recover, but it takes time. Some species might not return to the area for many years. In this case, the habitat can recover from the sudden change caused by the fire. But what happens when the changes are too extreme for an area to recover?

✏️ **Complete the exercise.**

Test your knowledge

(1) Which of these conditions would likely cause changes to a species' habitat?

A. the number of trees growing in an area

B. the number of rocks in a area

C. the amount of birds present

D. the amount of rainfall an area gets each year

Ans. ▢

(2) What happens to animals when their habitat changes quickly?

A. The animals always move north to find food and shelter.

B. The animals might move to other places.

C. All animals can adapt quickly to sudden changes.

D. All animals will become extinct.

Ans. ▢

Chapter 4
Habitat Change

Think!

How do people cause habitat change?

Check

 Read the key points below. When you finish, check the box.

Key Points: People and Habitat Change

People can also be the cause of changes to a habitat. One very visible way that humans change habitats is by changing the land. Busy highways are built through meadows. Forests are cut down to make way for houses and other buildings. Dams change the flow of rivers. All of these human actions can change habitats and that can put animal and plant populations in danger.

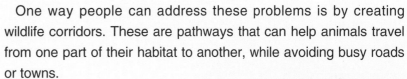

People are now becoming aware of their impact on the land around them and are taking steps to fix the changes they cause to some habitats.

One way people can address these problems is by creating wildlife corridors. These are pathways that can help animals travel from one part of their habitat to another, while avoiding busy roads or towns.

Another possible solution is to change how people design buildings. More buildings could have features that benefit local plants and animals. Once such feature, already being used on some buildings, is a green roof – a rooftop covered in plants. Scientists and engineers are continuing to work on other ideas for making buildings better for local species.

🖊 **Complete the exercise.**

Test your knowledge

Choose one of the best things you can do to prevent people from threatening animal and plant habitats.

A. Build a highway and drive many cars on it.
B. Build dams and change the flow of the river.
C. Cut down trees and build huge buildings.
D. Create a green area covered with plants on the roof of a building.

Ans. ☐

Chapter 4
Habitat Change

✏️ Use the word box below to fill in the blanks and review key vocabulary.

Review the Key Points

Every living thing is suited to a particular environment: its [   ]. A habitat is the natural home or environment of an animal, plant, or other living thing.

Changes to a habitat often happen slowly. For example, each year the temperature might rise a little each year or the amount of rainfall might drop.

Even if a habitat changes, species can sometimes [   ].

If changes to a habitat are extreme or happen quickly, plants and animals may not have enough time to adapt. Members of the species might have to move somewhere else, or, over time, they might adapt to the new conditions. If one of these things doesn't happen, the species could be come [   ] or die out.

[   ] can also be the cause of changes to a habitat. One very visible way that humans change habitats is by changing the land. Busy highways are built through meadows. Forests are cut down to make way for houses and other buildings. Dams change the flow of rivers. All of these human actions can change habitats and that can put animal and plant populations in danger.

One way people can address these problems is by creating wildlife [   ]. These are pathways that can help animals travel from one part of their habitat to another, while avoiding busy roads or towns.

> people / adapt / habitat / corridors / extinct

🖩 Complete the exercise.

Math Mission

There is a forest park A with an area of 4 km^2 and a forest park B with an area three times that of forest park A.

(1) Find the area of forest park B.

Ans. [   ] km^2

(2) The area of forest park C is equal to twice the total area of parks A and B. Find the area of the park C.

Ans. [   ] km^2

(3) Half of park C was destroyed by a forest fire. Find the area of the half of park C that did not burn.

Ans. [   ] km^2

Chapter 4
Habitat Change

Read the mission. Then, answer the following questions to help you with your solution.

The Mission

Imagine that you are an architect. Design a school building that does not damage the habitats of local plants and animals. You can plan to build the school near where you live, or anywhere else in the world.

Before you design...THINK!

1. Describe the mission in your own words.

2. Brainstorm your solution. Write your notes in the space below.

   Use the following questions to guide your thinking:

   (1) Where will you build your school? What kinds of plants and animals live there? What is their habitat like?

   (2) What are some features of your school building that would help preserve local habitats?

Habitat Change

💡 Read the mission. Then, draw and evaluate your solution.

The Mission

Imagine that you are an architect. Design a school building that does not damage the habitats of local plants and animals. You can plan to build the school near where you live, or anywhere else in the world.

Design   Draw or write about your solution below.

Evaluate

Evaluate your design. What challenges might students or teachers have in using your building? How could you improve your design to address those challenges?

Chapter 5
Invasive Species

 **Think!**

What is an invasive species?

 Check

 Read the key points. When you finish, check the box.

Key Points: What is an Invasive Species?

Cane toads, like the one shown here, don't have a lot of fans. They are larger than the average toad. Their skin is covered in warts. They eat nearly anything, including small animals. And they're poisonous. Pet owners have to be careful not to let their cat or dog catch one for a snack.

But to make matters worse, cane toads are spreading to new places. As they spread, they are making it harder for other plants and animals to survive.

Cane toads are an example of an **invasive species**. An invasive species is a species that does two things: It spreads to a new area, and it causes problems in that area.

Cane toads were once only found in Central and South America. Then, farmers brought the giant toads to other places such as Australia. The farmers wanted the toads to help eat bugs that were destroying crops.

But the toads reproduced (produced children) and spread faster than anyone expected. Now, the toads are found in much of Australia and other parts of the world. This is causing problems for **native species** in those places. A native species is a species that lives in a specific habitat.

Not all species that are new to an area are invasive. Some traits make a species more likely to be invasive. These traits include:
-It grows quickly.
-It reproduces quickly.
-It has no predators in the new area. (Nothing eats it.)

 Complete the exercise.

Test your knowledge

Choose the answer with the correct description of an invasive species.

A. Invasive species are species that have already been living in an area.

B. Invasive species spread to a new area, and they cause problems in that area.

C. Cane toads are now found only in Central and South America.

Ans.

Chapter 5
Invasive Species

Think!

Can you think of any invasive species in your area? What problems do they cause?

📖 Read the key points below. When you finish, check the box.
Check ✓

Key Points: Examples of Invasive Species

People often bring a species to a new place on purpose, like farmers did with the cane toad. Another example of this is a plant called kudzu that came from Japan to the United States. It was originally planted in the U.S. to help stop soil erosion (the process of soil washing or blowing away). Now, kudzu is an invasive species. Kudzu can grow up to a foot a day which can take nutrients away from native plants that grow around it.

kudzu

Sometimes, people accidentally bring a species to a new area, like the zebra mussel -- an invasive species in the Great Lakes of North America. These small animals were brought to North America on ships from Europe, without anyone realizing it. Zebra mussels are about the size of a fingernail and attach to hard surfaces. They can collect on and cover objects like boat parts, underwater pipes, and the shells of other animals.

zebra mussels

Another way invasive species come to a new area is as pets. This is how the Burmese python ended up becoming a problem in Florida. The snakes were brought from Asia and kept as pets. But, some of them escaped or were let go into the wild. Now, scientists think some native species in Florida are disappearing because pythons are eating them.

Burmese python

✏️ Complete the exercise.

Test your knowledge

Choose the appropriate invasive species based on the description of how it came to new areas.

> A. kudzu   B. zebra mussels   C. Burmese python

(1) A species that was kept as a pet and escaped or was released into new areas by people.

Ans. ☐

(2) A species that was accidentally brought to a new place by people.

Ans. ☐

(3) A species deliberately brought to a new place by people to stop soil erosion.

Ans. ☐

Chapter 5
Invasive Species

 Think!

What can happen to a native species if an invasive species changes its habitat too drastically?

 Read the key points below. When you finish, check the box.

Key Points: Problems caused by Invasive Species

Invasive species cause problems for native species in different ways. One way is simply by eating them. In Florida, Burmese pythons are feeding on native foxes and rabbits. The foxes and rabbits can't reproduce fast enough to maintain their population.

Another problem is that invasive species use a lot of resources such as food, water, and space. This means there are less resources available for native species. For example, kudzu grows in thick mats that take up space and block sunlight from reaching native plants.

In Chapter 4, you learned that every species lives in a certain environment, called a habitat. If an environment changes too drastically, native species might not be able to live there anymore. Invasive species can cause these kinds of changes. Zebra mussels, for instance, have caused changes in the amount of oxygen in the water where they live. Some native species cannot live in water with higher oxygen levels and are dying out.

Once an invasive species begins to take over, it can be really hard to get rid of. So, the best protection against invasive species is to make sure they don't arrive in the first place.

 Complete the exercise.

Test your knowledge

Match the invasive species with the problems each causes.

(1) kudzu  ●

● They are feeding on native foxes and rabbits.

(2) zebra mussels  ●

● Thick mats of them take up space and block sunlight from reaching native plants.

(3) Burmese pythons  ●

● They have caused changes in the amount of oxygen in the water where they live.

Chapter 5
Invasive Species

✏ Use the word box below to fill in the blanks and review key vocabulary.

Review the Key Points

An [ ] species is a species that does two things: It spreads to a new area, and it causes problems in that area.

People often bring a species to a new place on purpose. Sometimes, people accidentally bring a species to a new area. Another way invasive species come to a new area is as [ ].

Invasive species cause problems for [ ] species in different ways. One way is simply by eating them.

Another problem is that invasive species use a lot of resources such as food, water, and space. This means there are less [ ] available for native species.

If an environment changes too drastically, native species might not be able to [ ] there anymore. Invasive species can cause these kinds of changes.

native / live / resources / pets / invasive

⊞ Complete the exercise.

Math Mission

Kudzu is known to grow rapidly; assuming it grows a foot a day, answer the following questions.

(1) If the current length of the kudzu is 10 feet, how long will it be in 7 days?

Ans. [ ] feet

(2) How many days will it take for a 5 foot piece of kudzu to grow to 20 feet?

Ans. [ ] days

Chapter 5
Invasive Species

💡 **Read the mission. Then, answer the following questions to help you with your solution.**

The Mission

Think about one of the invasive species you learned about in this chapter or you know from your area. Design a plan to stop this invasive species from spreading farther.

For example, you might create a tool to help remove the species from the area. Or, you might plan a system to help people report when and where they see the species to better track it.

Before you design...THINK!

1. Describe the mission in your own words.

2. Brainstorm your solution. Write your notes in the space below.

Use the following questions to guide your thinking:

(1) What kind of invasive species are you going to focus on?
(2) What are some ways that this species might spread?

Chapter 5
Invasive Species

Read the mission. Then, draw and evaluate your solution.

The Mission

Think about one of the invasive species you learned about in this chapter or you know from your area. Design a plan to stop this invasive species from spreading farther.

Invasive Species Alert

Design

Draw or write about your solution below.

Evaluate

Evaluate your plan. What do you like best about your plan? What is one part of your plan that could be improved?

Chapter 6
Germs

Think!

Have your ever heard of bacteria? Do you know how bacteria can affect your body?

📖 Read the key points. When you finish, check the box.

Check ✓

Key Points: What are germs?

You know that one way to stay healthy is to wash your hands, because it gets rid of **germs** that could cause disease. But what, exactly, are germs?

Most germs are microbes, or tiny living things. They are too small to see with only your eyes. Not all microbes are harmful. There are plenty of microbes around us, and even in our bodies, that don't cause harm or disease. Some of them are even helpful.

However, for some of these microbes, it's a different story. If germs manage to get into your body, they can grow and reproduce (make more of themselves). They can then produce chemicals that are bad for your body or damage your cells. This can lead to disease. Germs can spread from person to person, which means the disease can spread too.

Germs can enter our bodies in different ways—when you touch your eyes or nose, or through a cut in your skin. But we can do a lot to keep germs away. Washing your hands is one important step. You may have seen a doctor or nurse wearing a mask. That is another way to prevent the spread of germs.

Science teaches us a lot about how germs live and grow. This helps us develop more ways to keep ourselves and our community healthy.

✏️ Complete the exercise.

Test your knowledge

(1) What are germs?
A. living things that eat cells
B. microbes or tiny living things
C. cells that live in human bodies
D. large living things that are dangerous

Ans. B

(2) How do germs enter the body?
A. through the mouth only
B. through your skin
C. through clothing
D. through openings in the body like the mouth or a cut

Ans. D

Chapter 6
Germs

Think!

What is bacteria? What is a virus? Let's think about the differences between the two.

 Read the key points below. When you finish, check the box.

Check

Key Points: Types of Germs

■ Most germs belong to one of two groups : **bacteria** or **viruses**.

Bacteria: Bacteria are living things made up of a single cell. This is very different from your body, which is made up of about 30 trillion (30,000,000,000,000) cells! Bacteria cells can be different shapes, including spheres, rods, and spirals.

Some kinds of bacteria cause disease. For example, you may have had a tetanus shot before. A tetanus shot protects you against a kind of bacteria that is found in dirt and soil. It can sometimes get into our bodies through a cut in the skin and make us sick.

Other kinds of bacteria are helpful. We use bacteria to make milk and cheese. You have bacteria living in your stomach that help you digest food. As you'll learn in Chapter 8, bacteria also help break down waste.

Viruses: If you've ever had a cold, you've had a virus in your body. Viruses are very different from bacteria—and from living things in general. Viruses are not made up of cells. They also cannot reproduce on their own. A virus has to use the cells of another living thing to reproduce. For example, when a cold virus infects your body, it takes over some of your cells to make more virus particles.

All germs are small, but viruses are really small. One virus particle is about 100 times smaller than just one bacteria cell.

Like bacteria, viruses are not all bad. Scientists are using some viruses to develop cures for serious diseases.

 Complete the exercise.

Test your knowledge

Answer T for true or F for false.

(1) All bacteria is bad for humans and can cause disease.

Ans. F

(2) One virus particle is larger than just one bacteria cell.

Ans. F

(3) Diseases can be caused by both bacteria and viruses.

Ans. T

Chapter 6
Germs

Can you think of ways to protect yourself from germs?

Check

Read the key points below. When you finish, check the box.

Key Points: Getting rid of Germs

■ We can get rid of germs in a lot of different ways. Some methods just clean germs off a surface. Other methods actually kill the germs.

Soap and Water: Normal soap and water doesn't kill germs. When you wash your hands, the soap loosens germs from your hands. Then, rinsing with water washes the germs off. Make sure to wash your hands for 20 seconds!

Chemicals: One way to kill germs is to use a chemical called a disinfectant. Many disinfectants can be dangerous and should only be used by or with an adult. Some common disinfectants are bleach, vinegar, and alcohol. You've probably used hand sanitizers that contain alcohol to kill germs on your hands before.

Heat: Very high temperatures can kill germs. That's why it's important to cook some foods, such as meat, before eating. However, the temperature of hot water you wash your hands in is not hot enough to kill germs. Hospitals often use very hot steam to clean germs from surfaces. Objects that need cleaning are put inside a special machine called a sterilizer. Hot steam is then safely applied inside the machine to kill germs.

UV Light: Some of the light from the sun and other sources is called ultraviolet (or UV) light. UV light can be harmful to your eyes and skin; that's why sunglasses and sunscreen are important when you're outside. However, lamps that produce UV light can be used to kill germs. Drinking water is sometimes treated with UV light to be sure it's safe to drink.

Complete the exercise.

Test your knowledge

Match the descriptions to the methods for getting rid of germs below.

A: soap and water

B: hand sanitizers (contain alcohol)

C: sterilizer

D: UV light

(1) This method kills germs with intense heat.

Ans. C

(2) This method kills germs with UV light or rays from the sun.

Ans. D

(3) This method kills germs with chemicals.

Ans. B

(4) This method rids your hands of germs, but does not kill them.

Ans. A

✏ Use the word box below to fill in the blanks and review key vocabulary.

Review the Key Points

Most [_____] are microbes, or tiny living things. They are too small to see with only your eyes. Not all microbes are harmful. There are plenty of microbes around us, and even in our bodies, that don't cause harm or disease. Some of them are even helpful.

Most germs belong to one of two groups : bacteria or viruses.

[_____] are living things made up of a single cell.   Bacteria cells can be different shapes, including spheres, rods, and spirals.

Some kinds of bacteria cause disease. Other kinds of bacteria are helpful. Bacteria also help break down waste.

[_____] are very different from bacteria— and from living things in general. Viruses are not made up of cells. They also cannot reproduce on their own. A virus has to use the cells of another living thing to reproduce.

Like bacteria, viruses are not all bad. Scientists are using some viruses to develop cures for serious [_____].

We can get rid of germs in a lot of different ways. Some methods just clean germs off a surface. Other methods actually kill the germs.

> viruses / bacteria / diseases / germs

🧮 Complete the exercise.

Math Mission

The graph on the right shows how the number of bacteria increases over time. Answer the following questions.

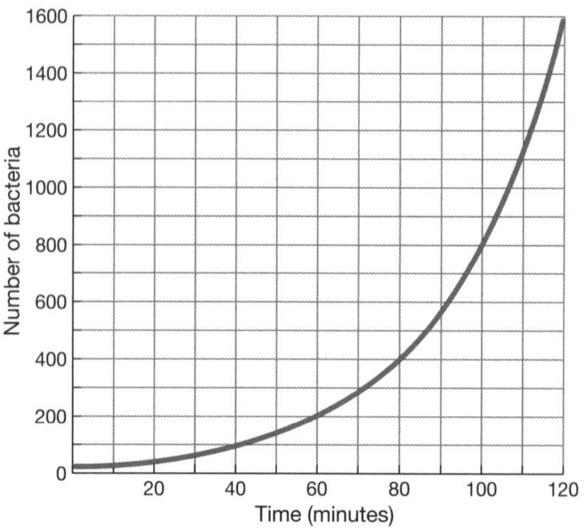

(1) How many bacteria are alive after 40 minutes?

Ans. [_____]

(2) How many bacteria are alive after 80 minutes?

Ans. [_____]

(3) How long did it take the number of bacteria to reach 1600?

Ans. [_____] minutes

Chapter 6
Germs

💡 Read the mission. Then, answer the following questions
  to help you with your solution.

The Mission

One way people can spread germs from place to place is on their shoes. This can be a problem in doctor's offices and hospitals, where people need to be extra careful about spreading germs. You've probably seen doctors and nurses wearing shoe covers to help solve this problem. But shoe covers have to be thrown away and create trash.

Your mission is to design a device that can quickly remove germs from people's shoes. Imagine your device would be used in hospitals, doctor's offices, and anywhere else where people need to be cautious about germs.

Before you design...THINK!

1. Describe the mission in your own words.

2. Brainstorm your solution. Write your notes in the space below.

   Use the following questions to guide your thinking:

   (1) What are some ways to kill germs or stop them from spreading?
   (2) What ways of getting rid of germs are safest for people? What ways are more dangerous?

Chapter 6
Germs

 Read the mission. Then, draw and evaluate your solution.

The Mission

Your mission is to design a device that can quickly remove germs from people's shoes. Imagine your device would be used in hospitals, doctor's offices, and anywhere else where people need to be cautious about germs.

Design

Draw or write about your solution below.

Evaluate

Evaluate your design. What challenges might people have in using your device? How could you improve your design to address those challenges?

Chapter 7
How Plants Grow

Think!

Plants need soil or dirt to grow. True or false?

 Read the key points. When you finish, check the box.

Check ✓

Key Points: How Plants Grow

If you take a look at the lettuce plants growing in this picture, you might notice that something is missing: Dirt. These plants are being grown without soil.

While it may look weird, there's nothing special about these lettuce plants. In facts, most plants can grow without soil.

We are used to seeing plants growing in fields or pots of dirt. But a plant doesn't actually need dirt to survive. Plants mainly need **water**, **sunlight**, and a gas called **carbon dioxide** from the air.

However, just like your body needs vitamins, plants also need a small amount of nutrients. Plants usually get these nutrients from the soil. But they can get them in other ways too.

People often think that when a plant is growing, it's using dirt to grow new leaves, stems, and flowers. But that is not true. The "stuff" that makes up a plant comes mainly from carbon dioxide and water. In some ways, plants seem like magic. They can practically grow out of thin air!

✏ Complete the exercise.

Test your knowledge

(1) What are the main things plants need to live and grow?
Choose three from the following.

A. water   B. sunlight

C. soil   D. carbon dioxide

Ans.

(2) Which of the following do plants need just like your body needs vitamins?

A. water   B. flowers

C. nutrients   D. oxygen

Ans.

Chapter 7
How Plants Grow

Think!

Do you know how plants make their food?

Check

📖 **Read the key points below. When you finish, check the box.**

Key Points: What is Photosynthesis?

Plants take in water through their roots. But a plant's leaves really hold the secret to how a plant grows.

Leaves capture energy from sunlight. They also take in carbon dioxide. Inside the leaves, the energy from sunlight is used to turn carbon dioxide and water into sugar. This process by which plants use carbon dioxide and sunlight to make sugar is called **photosynthesis**.

The sugar made by plants is very important. It is the plant's food, or source of energy. It's also a building block for all the parts of the plant—its leaves, stems, and roots. And what about when you eat part of a plant, such as a strawberry, carrot, or bean? What you are eating is made out of the sugars that the plant created.

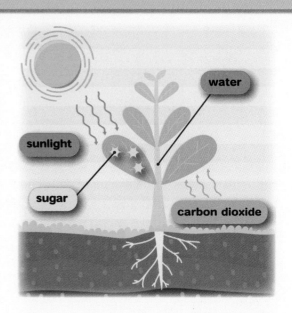

Plants do need nutrients to survive and be healthy. Some of these nutrients include iron and calcium. Plants take in these nutrients through their roots. Usually, the nutrients come from the soil where the plant is growing. You'll learn more about soil in Chapter 8. But when farmers grow plants without soil, like the lettuce plants at the beginning of this chapter, they can add nutrients to the water. This gives the plants everything they need, without the dirt.

✏️ Complete the exercise.

Test your knowledge

Choose one that correctly describes photosynthesis.

A. The process by which plants make sugar using oxygen and sunlight.

B. The process by which plants make sugar using carbon dioxide and sunlight.

C. The process by which plants make sugar using carbon dioxide and water.

D. The process by which animals make sugar using oxygen and sunlight.

Ans. ☐

Chapter 7
How Plants Grow

Think!

Did you know plants help humans grow? Read more to find out how!

 Read the key points below. When you finish, check the box.

Check

Key Points: How People and Plants Interact

You have learned that plants need carbon dioxide to make sugar. Where does this carbon dioxide come from? One source of this gas is you! Living things release carbon dioxide. You release it when you breathe out.

The reverse happens when you breathe in. When you inhale, you take in oxygen gas from the air. Where does this oxygen come from? Plants make it when they go through the process of photosynthesis.

When you look around you, at the air, it's hard to believe that anything is there. You can't see the carbon dioxide gas or the oxygen gas in the air with your eyes. That is one of the amazing things about living things, including ourselves. Living things use materials that we can't even see to live and grow.

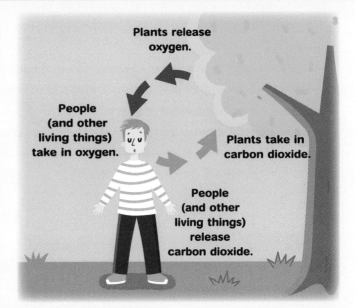

Plants release oxygen.

People (and other living things) take in oxygen.

Plants take in carbon dioxide.

People (and other living things) release carbon dioxide.

 Complete the exercise.

Test your knowledge

(1) Where does carbon dioxide come from?

A. non-living things   B. living things

C. the water   D. the soil

Ans. ☐

(2) Which is a product of photosynthesis?

A. air   B. oxygen

C. carbon dioxide   D. water

Ans. ☐

(3) Arrange the photosynthesis steps in order.

A. humans breathe out carbon dioxide

B. plants create sugar

C. plants take in carbon dioxide

D. plants release oxygen

Ans. $A \rightarrow \quad \rightarrow \quad \rightarrow$

Chapter 7
How Plants Grow

 Use the word box below to fill in the blanks and review key vocabulary.

Review the Key Points

We are used to seeing plants growing in fields or pots of dirt. But a plant doesn't actually need dirt to survive. Plants mainly need [   ], sunlight, and a gas called [   ] from the air.

However, just like your body needs vitamins, plants also need a small amount of nutrients. Plants usually get these nutrients from the soil. But they can get them in other ways too.

Leaves capture energy from sunlight. They also take in carbon dioxide. Inside the leaves, the energy from sunlight is used to turn carbon dioxide and water into sugar. This process by which plants use carbon dioxide and sunlight to make sugar is called [   ].

The reverse happens when you breathe in. When you inhale, you take in [   ] gas from the air. Where does this oxygen come from? Plants make it when they go through the process of photosynthesis.

> oxygen / carbon dioxide / water / photosynthesis

Complete the exercise.

Math Mission

You have learned that one of the things that plants need to survive is sunlight.

The table on the right shows the times when the sun rises and sets at three locations A, B, and C on the same day.

	sunrise	sunset
location A	6:45	18:45
location B	6:30	18:50
location C	6:55	19:05

(1) In location A, find the time from sunrise to sunset.

Ans. [   ] hours

(2) How much longer is the time from sunrise to sunset in location B compared to location A?

Ans. [   ] minutes

(3) Where is the location with the longest time from sunrise to sunset?

Ans. [   ]

How Plants Grow

 Read the mission. Then, answer the following questions to help you with your solution.

The Mission

Many people dream that one day, people will be able to travel long distances through space. One thing standing in the way of that dream is having enough food. We can't bring all the food we would need with us. We would have to grow it.

Your mission is to design a spaceship garden. How would you grow plants on a spaceship? How would you make sure the plants get everything they need?

Before you design...THINK!

1. Describe the mission in your own words.

2. Brainstorm your solution. Write your notes in the space below.

   Use the following questions to guide your thinking:

   (1) What do plants need to survive and grow?
   (2) How would you meet each of these needs on a spaceship?

Chapter 7
How Plants Grow

Read the mission. Then, draw and evaluate your solution.

The Mission

Your mission is to design a spaceship garden. How would you grow plants on a spaceship? How would you make sure the plants get everything they need?

Design   Draw or write about your solution below.

Evaluate

Evaluate your garden. Do you think it could grow enough food? How could you make it grow even more food? Are you missing any key components for growing plants?

Chapter 8
Decomposition

Think! Why are decomposers and decomposition important for the earth?

Check✓

📖 Read the key points. When you finish, check the box.

Key Points: What is Decomposition?

For most of us, an empty banana peel or brown apple core means that snack time is over. But for a **decomposer**, the meal is just getting started. Decomposers are living things that feed on dead plant and animal parts. Some examples of decomposers are mushrooms, worms, insects, and bacteria. The process that decomposers carry out is called **decomposition**.

When decomposers feed, they break down plant and animal matter into nutrients. This is decomposition. They then release these nutrients back into the environment, so other living things can then use them. For example, some of the nutrients released by decomposers end up in soil. Nutrients in the soil can be used by plants, as you learned in Chapter 7. People often call decomposers "nature's recyclers" because they allow important nutrients to get reused.

We rely on decomposers every day. They help us get rid of waste. This can include food waste, like you read about above. It also includes things like dead and decaying material in forests, and even animal waste. Without decomposers, we'd be surrounded by waste, and plants could not get the nutrients they need to live and grow.

✏️ Complete the exercise.

Test your knowledge

(1) What is decomposition?
A. the natural process of plant growth
B. how plants get nutrients to grow
C. the process through which plants reproduce
D. the breakdown of plant and animal matter into nutrients

Ans. ☐

(2) What is a decomposer?
A. a living thing that reproduces slowly
B. a non-living thing that reproduces slowly
C. a living thing that feeds on dead plant and animal parts
D. a non-living thing in the soil

Ans. ☐

Decomposition

 **Think!**

What types of decomposers have you seen at work?

 Read the key points below. When you finish, check the box. ✓

Check

Key Points: Types of Decomposers

Decomposers can be found anywhere there is waste material for them to break down. One place you'll always find them is outside. You can find different decomposers in the soil, on trees, or under dead leaves. However, some kinds of decomposers are too small to see with your eyes.

Three major types of decomposers are **earthworms**, **fungi** (or fungus), and **bacteria**.

Earthworms: You've probably flipped over a rock or log and seen an earthworm below it. Earthworms are a kind of animal called an invertebrate. Invertebrates do not have bones. Earthworms eat dead plant or animal material in the soil and release nutrients in their waste, called castings. Castings are a great source of nutrients for growing plants. Earthworms also create tunnels through the soil. This benefits other living things in the soil by helping move air and water around.

Fungi (or Fungus): One familiar kind of fungus are mushrooms. Some mushrooms look a little like plants. But mushrooms don't need sunlight to grow. They get all the energy and nutrients they need from decomposing plants.

Another type of fungi is mold. If you let a loaf of bread sit around for too long, you might notice greenish mold growing on it. The mold is decomposing the bread.

Bacteria: Remember that you read about bacteria in Chapter 6. Some, but not all, bacteria can cause disease. Other kinds of bacteria are helpful. Have you ever taken a deep breath outdoors and noticed that the dirt smells "fresh"? That fresh smell comes from a certain kind of bacteria that breaks down waste material into soil. It's a sign that the nutrient levels in the soil are healthy. Just a spoonful of dirt might contain up to 1 billion (1,000,000,000) bacteria!

Complete the exercise.

Test your knowledge

Identify the decomposer based on the descriptions below.

> **A. Earthworms   B. Fungi (or Fungus)   C. Bacteria**

(1) This decomposer is an invertebrate that moves through soil and leaves castings in the soil that provide nutrients for plants.

Ans. ☐

(2) This decomposer is too small to be seen by the human eye. It lives in the soil with billions of its friends. Together they breakdown plant and animal waste material into nutrients for the soil.

Ans. ☐

(3) This decomposer might look like a plant, but it is not. It gets its food from breaking down waste materials. It also does not need sunlight to grow. It gets all of its energy from decomposing waste materials.

Ans. ☐

Chapter 8
Decomposition

Think!

Do you think people can be decomposers?

Check

📖 Read the key points below. When you finish, check the box. ✓

Key Points: What is Compost?

You probably already know how important it is to recycle. We can recycle materials such as paper, plastics, and metals. Recycled materials are used to make new things so they don't end up as trash.

There is also a way to recycle nutrients from food waste and other materials. This process is called making compost. To make **compost**, you collect certain kinds of waste and let decomposers break it down. Eventually a dirt-like material forms—this is the compost. When you add compost to soil, plants and other living things can use the nutrients in it. In this way, the nutrients get recycled.

Anyone can make compost with enough space and the right materials. The basic steps to follow when making compost are:

Step 1:
Collect a mix of natural waste materials (mostly from plants).

Step 2:
Add decomposers and make sure they have what they need to live. This includes the right temperature, water, and amount of air.

Step 3:
Keep repeating these steps and give the decomposers time to make compost! This can take several months.

Although composting follows the same basic steps, people use different systems. Some people can just create the compost in a pile, or they will use a bin. People might need to mix up the waste, either by hand or by turning the bin to add more air. They can even choose whether or not to add worms to their compost.

✏️ Complete the exercise.

Test your knowledge

Choose the best explanation for Composting.
A. the process of collecting waste materials and breaking them down into nutrients
B. the process of recycling plastic or glass into something new
C. the process of preventing soil erosion
D. the process of creating a new habitat for plants

Ans.

Chapter 8
Decomposition

✏ Use the word box below to fill in the blanks and review key vocabulary.

Review the Key Points

Decomposers are living things that feed on dead plant and animal parts. Some examples of decomposers are mushrooms, worms, insects, and bacteria.

When decomposers feed, they break down plant and animal matter into nutrients. This is [_____]. They then release these nutrients back into the environment, so other living things can then use them.

Decomposers can be found anywhere there is waste material for them to break down. One place you'll always find them is outside. You can find different decomposers in the soil, on trees, or under dead leaves. However, some kinds of decomposers are too small to see with your eyes.

Three major types of decomposers are earthworms, [_____] (or fungus), and [_____].

There is also a way to recycle nutrients from food waste and other materials. This process is called making compost. To make [_____], you collect certain kinds of waste and let decomposers break it down. Eventually a dirt-like material forms—this is the compost. When you add compost to soil, plants and other living things can use the nutrients in it. In this way, the nutrients get recycled.

> **fungi / decomposition / compost / bacteria**

🖩 Complete the exercise.

Math Mission

Use the chart to answer the questions about composting.

Total Amount of Trash Wasted Compared to the Total Amount of Trash That Could Be Composted

Legend:
- Total Amount of Trash
- Total Amount of Trash That Could be Composted

(1) Find the "Total Amount of Trash" on Day 5.

Ans. [_____] gallons

(2) Find the "Total Amount of Trash That Could be Composted" on Day 3.

Ans. [_____] gallons

(3) Find the difference between "Total Amount of Trash" and "Total Amount of Trash That Could be Composted" on day 9.

Ans. [_____] gallons

Chapter 8
Decomposition

 Read the mission. Then, answer the following questions to help you with your solution.

The Mission

Design a system for composting food scraps. It can be for your home, your school, or somewhere else in your community. Then, be sure to point out how your system will encourage people to compost.
In other words, for this mission you'll want to think about two things:
1) How compost is made.
2) How to get people to participate.

Before you design...THINK!

1. Describe the mission in your own words.

2. Brainstorm your solution. Write your notes in the space below.
   Use the following questions to guide your thinking:

(1) What are the steps for making compost? What type of decomposers will you use?
(2) Think of some reasons people might not like to compost. How can you make your system better?

Decomposition

💡 Read the mission. Then, draw and evaluate your solution.

The Mission

Design a system for composting food scraps. It can be for home, your school, or somewhere else in your community. Then, be sure to point out how your system will encourage people to compost.

Design   Draw or write about your solution below.

Evaluate

Evaluate your composting system. Would you want to use it? Why or why not?

Chapter 1 **Plant Adaptations**

1

Test your knowledge

(1) C   (2) B

2

Test your knowledge

(1) C   (2) A   (3) B

3

Test your knowledge

(1) B   (3) A   (5) A

(2) C   (4) B   (6) D

4

Review the Key Points

adaptation / environment / seeds / animals

Math Mission

(1) 50 mm   (2) A   (3) B

5 (Sample Response)

Before you design... THINK!

1. Create a new type of seed that will travel far, so it can find a good place to grow.
2. My seed would use an exploding seed pod and smooth seed shape to sail through the air and land farther away from the parent plant. It would grow in a field like environment, so good soil is nearby.

6 (Sample Response)

Design

My seed would explode from a seed pod. It would be streamline and smooth, like an arrowhead. So it would sail through the air further from the parent plant.

Evaluate

My seed looks like a sunflower's seed, but animals cannot eat it. So it needs to explode from the pod and fly through the air. As long as it finds good soil it will grow.

Chapter 2 **Pollination**

7

Test your knowledge

(1) A   (2) B

8

Test your knowledge

(1) C   (2) D

9

Test your knowledge

(1)
(2)
(3)

10

Review the Key Points

Pollination / pollinator / pollen

Math Mission

(1) $25 \times 10 = 250$ Ans. 250 flowers

(2) $1000 \times 5 = 5000$ Ans. 5000 flowers

(3) $12 \div 4 = 3$ $2 \times 3 = 6$ Ans. 6 hours

11 (Sample Response)

Before you design... THINK!

1. Create a tool to help farmers pollinate more flowers on their apple trees.
2. Bees are good pollinators because they have tiny hairs on their bodies that collect and hold pollen. They can also fly from one flower to another, which helps cross-pollinate many flowers.

Other traits from pollinators that could be helpful would be the shape of their mouths or body parts that collect the pollen. Like a hummingbird's beak helps it get food and pollen from long tube like flowers.

12 (Sample Response)

Design

- Use a stick with pipe cleaners attached. They can be bent in different ways to reflect different types of flowers / pollinaters.
- The bristles will mimic the hair on bee's legs.

Evaluate

One part of my device that could be improved is how the pollen is removed from the bristles. With this device the farmer has to shake the stick over the flowers to get the pollen off. I think there could be a better method.

Chapter 3 **Animal Adaptations**

13

Test your knowledge

(1) •
(2) •
(3) •

14

Test your knowledge

A → E → B → D → C

15

Test your knowledge

(1) •
(2) •
(3) •
(4) •
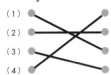

16

Review the Key Points

adaptations / food / child / beaks

Math Mission

(1) 4   (2) 10   (3) spoon

17 (Sample Response)

Before you design... THINK!

1. Create an animal that could live in the cave environment.

2. The weather might be warm during the day and cold at night because the cave is dark.

   Small plants would grow here because it is dark and only gets a little sunlight.

   Being able to see in the dark would be useful to animals who live in this type of environment.

18 (Sample Response)

Design

My animal would have big eyes so it could see better in the low light of the cave. It would have a sticky tongue like a frog to help it catch the bugs and lizards that live in the cave too. My animal would have wide paddle-like feet like a duck to help it walk on the muddy ground.

Evaluate

I think my animal could be better adapted to surviving in the wet cave environment if it had thick fur to keep it warm. An adaptation that would not help my animal would be sharp claws on its feet because they would sink into the mud.

Chapter 4 **Habitat Change**

19

Test your knowledge

(1) •
(2) •
(3) •
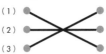

20

Test your knowledge

(1) D   (2) B

21

Test your knowledge

D

22

Review the Key Points

habitat / adapt / extinct / People / corridors

Math Mission

(1) $4 \times 3 = 12$ Ans. 12 km^2
(2) $(4 + 12) \times 2 = 32$ Ans. 32 km^2
(3) $32 \div 2 = 16$ Ans. 16 km^2

23 (Sample Response)

Before you design... THINK!

1. Build a school building that is safe for animals and people.

2. I could build my school building off the side of a mountain or hill. I would make sure that it would not bother the trees growing there. My building would be built around trees, so no land has to change.

24 (Sample Response)

Design

Instead of leveling a hill side to build on flat ground, we can build a platform off the side to support the school not change the landscape.

Evaluate

One challenge is it would take a lot of resources to build the platform to support a whole building. The land would still be changed by the supports.

Chapter 5 Invasive Species

25

Test your knowledge

B

26

Test your knowledge

(1) C   (2) B   (3) A

27

Test your knowledge

(1)
(2)
(3)

28

Review the Key Points

invasive / pets / native / resources / live

Math Mission

(1) 17 feet   (2) 15 days

29 (Sample Response)

Before you design... THINK!

1. Create a plan to help stop invasive animals or plants from finding their way to new places.
2. I think I will focus on zebra mussels and how to stop them from moving to new bodies of water.
   Since they spread using boats, I will find a way to stop them from getting on the boats in the first place.

30 (Sample Response)

Design

- Create a spray for the hull of boats that is slippery and doesn't allow the mussels to attach to the boats.
- This will help stop people from bringing them by boat to other waterways.

Evaluate

What could be improved is how to get people to use the spray. Can it be made mandatory for people use? We would also have to make the spray safe for the environment so it doesn't cause more problems.

Chapter 6 Germs

31

Test your knowledge

(1) B   (2) D

32

Test your knowledge

(1) F   (2) F   (3) T

33

Test your knowledge

(1) C   (2) D   (3) B   (4) A

34

Review the Key Points

germs / Bacteria / Viruses / diseases

Math Mission

(1) 100   (2) 400   (3) 120 minutes

35 (Sample Response)

Before you design... THINK!

1. Create a way to kill germs, or prevent them from spreading, that is safe for people.
2. Some ways to kill germs are: UV light, super hot steam, or chemicals.
   We can also stop germs from spreading by washing our hands with soap and water.
   Soap and water is safe for people to use to get rid of germs.

36 (Sample Response)

Design

- Stand on UV lightbulbs for 1 minute.
- Kill germs on shoes.

Evaluate

One challenge would be that UV light can be harmful to humans, the device would have to only shine light on the person's shoes.

Chapter 7 How Plants Grow

37

Test your knowledge

(1) A, B, D   (2) C

38

Test your knowledge

B

39

Test your knowledge

(1) B   (2) B   (3) A → C → B → D

40

Review the Key Points

water / carbon dioxide / photosynthesis / oxygen

Math Mission

(1) 12 hours   (2) 20 minutes   (3) B

41 (Sample Response)

Before you design... THINK!

1. Create a plan for growing plants without soil on a spaceship.
2. Plants need water, nutrients, carbon dioxide, and sunlight to grow.
   People produce carbon dioxide so we would not need to bring it with us. People also need water to drink so we would have to bring a lot of water for people and to grow plants.
   We would also need to make sunlight and provide the nutrients for the plants.

42 (Sample Response)

Design

- Rotating wheel so plants get equal sunlight and carbon dioxide as they move.
- Create a light bulb that acts as artificial sunlight to provide the plants with sun.
- Use recycled water from people cooking and cleaning on the spaceship to water the plants.

Evaluate

The method could grow a lot of food. However it would be hard to bring all the liquid nutrients to feed the plants.

Chapter 8 Decomposition

43

Test your knowledge

(1) D   (2) C

44

Test your knowledge

(1) A   (2) C   (3) B

45

Test your knowledge

A

46

Review the Key Points

decomposition / fungi / bacteria / compost

Math Mission

(1) 4000 gallons
(2) 1200 gallons
(3) 3600 gallons

47 (Sample Response)

Before you design... THINK!

1. Create a plan to make people compost more.
2. The steps for composting are compiling the waste materials, adding a decomposer, and making sure the compost has a safe place to decay.
   I could use worms as decomposers.
   Some people might not like compost in their homes because it is smelly. Other people might not have the space for it.

48 (Sample Response)

Design

- Create a town incentive program for bringing waste materials to a local compost site.
- Offer rewards for every bag of food scraps saved and composted. For example, if people bring a pound of food scraps to add to the compost they can take home fruit and vegetables grown in the compost garden.

Evaluate

This would be a good system, but it would take time to get people to use it. It would also cost a lot to reward people for participating.